I0005998

No. XXIX

Automata Old and New

BY

CONRAD WILLIAM COOKE, M.INST.E.E.

Mechanick to the Sette of Odd Volumes

LONDON

IMPRINTED AT THE CHISWICK PRESS

MDCCCXCIII

Automata Old and New

BY

CONRAD WILLIAM COOKE, M.INST.E.E.

Reprinted from the Sette of Odd Volumes

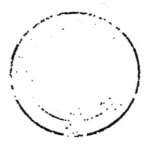

LONDON

PRINTED AT THE CHISWICK PRESS

M DCCC XCI

PRIVATELY PRINTED OPUSCULA.

ISSUED TO MEMBERS OF THE SETTE OF ODD VOLUMES.

No. XXIX. ·

AUTOMATA OLD AND NEW.

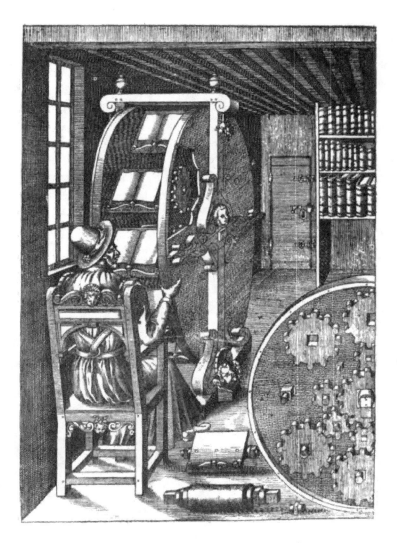

[*See page* 54.

Automata Old and New

BY

CONRAD WILLIAM COOKE, M.INST.E.E.

Mechanick to the Sette of
Odd Volumes

*Delivered at a Meeting of the Sette held at
Limmer's Hotel, on Friday,
November 6th,* 1891

LONDON

IMPRINTED AT THE CHISWICK PRESS
MDCCCXCIII

Eng 1838.93,5

To their Oddships

CHARLES HOLME, F.L.S.

(*Pilgrim*),

President, 1890.

GEORGE CHARLES HAITÉ, R.B.A., F.L.S.

(*Art Critic*),

President, 1891.

AND

WILLIAM MURRELL, M.D.

(*Leech*),

President, 1892.

DURING WHOSE YEARS OF OFFICE

THE FOLLOWING NOTES ON

AUTOMATA

WERE RESPECTIVELY

PREPARED, PRESENTED AND PRINTED,

THIS LITTLE BOOK

IS DEDICATED BY

Conrad W. Cooke,

Mechanick to ye Sette of Odd Volumes.

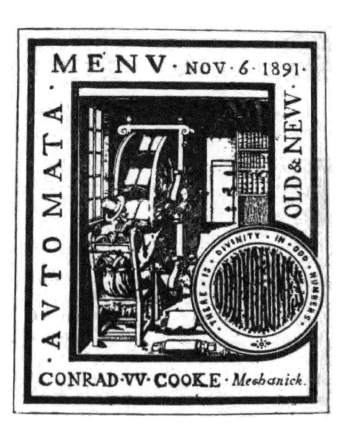

MENV · NOV · 6 · 1891 ·

AVTOMATA ·

OLD & NEW

THERE · IS · DIVINITY · IN · ODD · NUMBERS

CONRAD · VV · COOKE · *Mechanick.*

AUTOMATA OLD AND NEW.

AY it please your Oddship, Brethren and Guests of Yᵉ Sette of Odd Volumes. The origin of this little paper is very simple. Just eleven months ago we had the delight of listening to the very interesting and instructive communication upon the work of that wonderful mechanical genius, electrician, and *prestidigitateur*, Robert-Houdin, presented to us by my very good friend, our revered Seer, Brother Manning. With the object of contributing something to the discussion which followed that paper, I began to make a few notes upon Automata, with which subject the name of Robert-Houdin must for ever be associated; I soon found, however, that

the subject was so comprehensive and went
back into such remote periods of antiquity, that
to do it even the most scanty justice would
require a paper devoted to itself alone; and,
as our esteemed Pilgrim and Past-President,
Brother Holme, was at that time pressing me
for a paper with that persistency and impor-
tunity which characterized his presidentship and
gave it so much of its success, I, as a loyal Odd
Volume, felt bound to obey the mandate of
his Oddship; and, holding the honourable office
of *Mechanick* to the Sette, I have chosen "Automata Old and New" for the subject of this
communication.

The word Automaton would in its strictest and
most comprehensive sense include all apparently
self-moving machines or devices which contain
within themselves their own motive power, and
in this sense such machines as clocks and
watches, and even locomotives and steamships
might be included. I shall, however, through-
out this paper limit myself to the more restricted

and more ordinarily accepted meaning of the term, namely, such self-moving machines as are made either in the forms of men or of animals, or by which animal motions and functions are more or less imitated.

As mechanics, next to mathematics and astronomy, is the most ancient of sciences, and as the scientific knowledge of the ancients was ever shrouded in mystery to conceal it from the eyes of the vulgar, and to confer upon the initiated power and profit by working on the credulity of the ignorant, it was but only to be expected that mechanical science should be early applied in the ancient mysteries by which the philosophers and the priests of antiquity maintained so much of their supremacy.

One of the very earliest allusions to mysterious self-moving machines is to be found in the eighteenth book of the "Iliad," wherein we are told of Vulcan that

" Full twenty tripods for his hall he fram'd,
That, placed on living wheels of massy gold

(Wondrous to tell) instinct with spirit roll'd
From place to place, around the bless'd abodes,
Self-mov'd, obedient to the beck of gods."[1]

Several others of the ancient poets besides Homer have sung about the wonderful mechanical devices of Vulcan, among which were golden statues, the semblances of living maids, which not only appeared to be endued with life, but which walked by his side and bore him up as he walked. Aristotle also refers to self-moving tripods, and Philostratus states that Appolonius of Tyana saw similar pieces of mechanism among the Brahmins of India; but this must have been nearly four hundred years after Aristotle wrote, and some nine hundred years after the time of Homer.

Then again we hear of Dædalus making self-moving statues, small figures of the gods, of which Plato in his "Menos" says that unless they were fastened they would of themselves

[1] The "Iliad" of Homer, translated by Alexander Pope, xviii. 440-444.

run away, and he puts this into the mouth of
Socrates, who uses it as a figure to illustrate the
importance of not only acquiring but of holding
fast scientific truth that it may not fly away
from us. Aristotle in referring to these statues
affirms that Dædalus accomplished his object
by putting into them quicksilver, but the learned
mechanician Bishop Wilkins points out that
"this would have been too grosse a way for so
excellent an artificer; it is more likely that he
did it with wheels and weights."[1] We are moreover told by Macrobius[2] that in the temple of
Hieropolis at Antium there were moving statues.

A contemporary of Plato and, it is said, his
master, was Archytas of Tarentum, the celebrated Pythagorean philosopher, mathematician,
cosmographer, and mechanician, to whom is
accredited the invention of the screw and of the

[1] "*Mathematicall Magick*, or the Wonders that may
be performed by Mechanicall Geometry." London,
printed by *M. E.* for *Sa: Gellibrand* at the Brasen
Serpent in *Paul's* Church-yard, 1648 (page 173).

[2] "Saturnaliorum Conviviorum," Lib. I. cap. xxiii.

crane. Archytas is said to have constructed of
wood a pigeon that could fly about, but which
could not rise again after it had settled; and
Aulus Gellius (who lived in the reigns of
Trajan, Hadrian, Antoninus Pius, and Marcus
Aurelius), tells us in his "Noctes Atticæ," that
"many men of eminence among the Greeks,
and Favonius, the philosopher, a most vigilant
searcher into antiquity, have in a most positive
manner assured us that the model of a pigeon,
formed in wood by Archytas, was so contrived
as by a certain mechanical art and power to fly;
so nicely was it balanced by weights and put in
motion by hidden and inclosed air. In a matter
so very improbable we may be allowed to add
the words of Favonius himself: 'Archytas of
Tarentum, being both a philosopher and skilled
in mechanics, made a wooden pigeon which had
it ever settled would not have risen again till
now.'"[1] And I am bound to admit that in this
point I agree with him.

[1] Aulus Gellius, "Noctes Atticæ." Lib. X. cap. xii.

From the above description it would appear that a still greater invention than a flying automaton was made by Archytas, for in an apparatus " *so nicely balanced by weights and put in motion by hidden and inclosed air,*" we have a very fair forecast of the modern aërostat or balloon, filled with gas and balanced by ballast. There cannot be any doubt but that the accounts of these very early machines (if such ever existed at all), have been greatly exaggerated during the process of being handed down through long ages of ignorance and credulity; but we may now enter upon surer ground although still very ancient. In the reign of Ptolemy Euergetes II. (Ptolemy VII.), about 150 years B.C., there lived at Alexandria that great genius of mechanical science, Hero; and his remarkable book " Spiritalia," of which I am able to show you several copies to-night, is itself a great storehouse of ingenuity in the construction of automata of very various forms and principles. This remarkable man was, if not the inventor, the first

B

describer of the siphon in both its typical forms,
the syringe, the well-known portable shower-
bath, the clack valve, the fire engine, even with
that mechanical refinement, an air vessel for
insuring a continuous stream, a self-trimming
lamp, the steam blowpipe, the pneumatic foun-
tain called after his name, a steam engine, and
last if not least, the penny-in-the-slot automatic
machine for obtaining a drink, or, may be, a
charge of scent.

I propose now to show you on the screen
some photographic reproductions of pages in his
book, some taken from the Latin edition of
Commandinus, published at Urbino in 1575,
and some from the Italian edition of Alessandro
Georgi, printed at the same place in 1592, some
from the fine edition of Aleotti, published in
1589, and others from the Amsterdam version
of 1680, all of which editions I am able to show
you.   I have, moreover, copied some from
manuscripts in the British Museum, of the fifteenth and sixteenth centuries, of which there

are four in the National Library, *i.e.*, two in the Harley Collection and two among the Burney manuscripts.

The first illustration I shall show you from Hero's work is a bird which, by means of a

Fig. 1.

stream of water, is caused to pipe or sing. This little automaton consists of a pedestal (A B C D) (Fig. 1), which is in reality a water-tight tank fitted with a funnel (E), the stem of which reaches nearly to the bottom; to the right of this there is a little bush on which sits a bird, and a tube

(G H) leads up from the roof of the tank and terminates in a little whistle, the end of which dips into a cup (L) containing water. When water is poured into the funnel, the air in the tank is driven out through the tube and whistle (G H) and, bubbling through the water, sounds as if the bird were singing. Thus the well-known bubbling bird-whistle dates back to a century and a half before the Christian era or earlier.

The next illustration (Fig. 2) shows a more elaborate arrangement, in which there are four small birds being watched by an owl; the moment the owl's back is turned the birds begin to sing, but cease as soon as he turns towards them. In this apparatus the birds are made to sing in precisely the same way as in the last illustration, namely, by the displacement by water of the air in the tank, but as soon as the level of the water in the tank reaches the top of a concentric siphon (F G) the water is discharged into a bucket, the birds cease to sing, and the bucket, owing to its increased weight, lifts the counterbalance weight (z), and in doing so turns the spindle (p m) which supports the owl (r s). When the bucket is full its contents are discharged by a small siphon within it and it

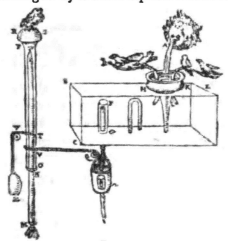

Fig. 2.

is drawn up by the weight (z) the owl turns its back to the birds, and the cycle of operations is repeated.

In the next figure a still more elaborate effect is produced. Here is a pedestal upon which are

four little bushes each having a bird sitting in its
branches ; when water is allowed to flow into the

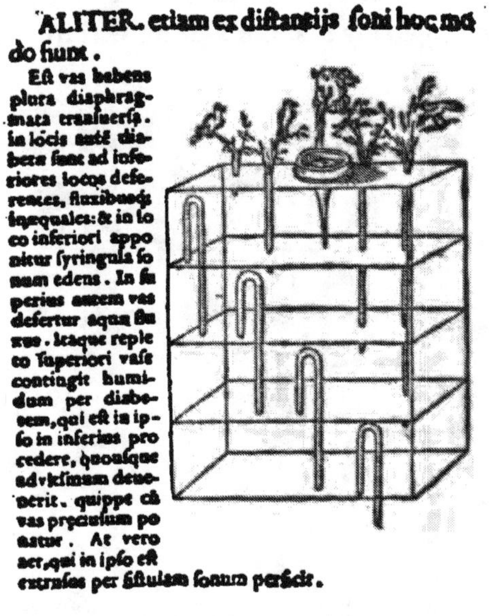

Fig. 3.

funnel the first bird begins to whistle, and after
a few minutes leaves off, when the next bird
begins, and when he has finished the third bird

sings, after a little time the fourth takes up the song, and when he has finished the first begins again, and so on as long as water is flowing into the funnel. These effects are pro-

Fig. 4.

duced in the simplest possible manner, by a com-bination of as many superposed tanks as there are birds to sing, the one emptying into the other by siphons. The illustration explains itself.

In the next device (Fig. 4) we have a bird whose singing is *intermittent*. In this case the water flows into a little cup which topples over the moment it is full, emptying itself into the funnel and immediately righting itself (being loaded at its bottom), the sound is produced by the displaced air escaping through a whistle in the manner already described.

We now come to a different class, in which heat is employed for obtaining an increase of air pressure whereby certain automatic actions are produced. Here we have a priest and priestess officiating at an altar; and the effect of lighting the fire thereon is to cause the two figures to pour libations onto the sacrifice. In this case the altar consists of an air-tight metallic box in communication, by means of a central tube, with a larger box forming the pedestal. Into this lower reservoir is poured the wine or other liquid through the hole marked M. When the fire is lighted the air in the altar is expanded, and pressing on the surface of the liquid in the pedestal, forces some of it through the tubes which pass through the body and down the right

arm of each figure. In the next view (Fig. 6)
we see how this principle was employed by
Hero for the opening of the doors of a temple,

Fig. 5.

the tradition being that when a sacrifice was
offered on her altar the goddess Isis showed her
invisible presence by throwing open the doors
of her sanctuary. In this case the altar consists
of an air-tight metallic box communicating by

means of a tube (F G) with a spherical vessel
(H) partly filled with water. When the altar
becomes hot the contained air is expanded, thereby increasing the pressure on the surface of the
water, some of which is therefore forced through

Fig. 6.

the bent tube (L) into the bucket (M), which
descends by its increased weight, thereby un-
winding the cords from the two spindles that
perform the function of hinges to the temple
doors, at the same time winding up the counterweight (R) on the left. When the fire goes out

the altar cools, assuming its ordinary atmospheric pressure, and the water in the bucket is forced back into the vessel (H), and the weight counterbalancing the empty bucket, closes again the doors.

Like many other geniuses who have lived before their time, Hero had his plagiarists, his devices having been adopted and described by later writers without one word of acknowledgment as to their authorship. From the middle to the end of the seventeenth century several books appeared which to a great extent were simply bad and erroneous copies of Hero's inventions, and not even intelligently copied. Here for instance (Fig. 7) is a *facsimile* of an illustration in a curious old book, "The Mysteries of Nature and Art," by John Bate, published in 1635; this is poor Bate's attempt to steal Hero's device for the temple doors, showing an altogether impossible scheme. In the first place the doors could not open at all, for the ropes are so coiled as to neutralize each other's ac-

tion, and, secondly, the counterweight to the
right has its cord simply looped round the
spindle and therefore is absolutely useless; the
accompanying description is even more absurd,

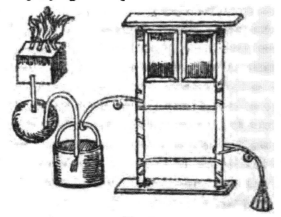

Fig. 7.

for it explains the action of the apparatus as
follows: "The fier on the Altar will cause the
water to distill out of the Ball into the Bucket,
which when (by reason of the water) it is be-
come heavier than the waight, it will draw it up
and so open the sayd gates or little doores."

Again, in one of Hero's illustrations a revolving disc carrying little figures was made to rotate upon the reaction principle of his own Æolipile, or steam engine. By a little bit of bad perspective the ends of the cross tubes were shown as turning alternately up and down, and Bate not only repeats this error, but goes out of his way to point out that "in the middest" there must be "a hollow pipe spreading itself into foure severall branches at the bottom : *the ends of two of the branches must turn up and the ends of two must turn down,*" thus making any rotative action impossible.

But Bate was not the only pirate of Hero's work ; a few years after Bate had written, that is, in 1659, there appeared another curious book by Isaak de Caus, upon Water Works,[1] and in

---

[1] "*New and Rare Inventions of Water Workes,* shewing the easiest waies to raise water higher than the spring. By which invention the Perpetual Motion is proposed, many hard labours performed And variety of Motions and Sounds produced. First written in French by Isaak de Caus a late famous engineer ; and now translated into English by John Leak. London, Printed by Joseph Moxon. Folio. 1659."

that book we find our old friend the owl keeping the small birds in order, the only difference

est autem constructio talis.

Sit vas A B, in
quod influit cana
liculus C; & in eo
fit inflexus fipbô,
fine fpiratilis dia
betes D E F, cu-
ius alterum crus
extra vafis fundû
excedat. ipfi vero
fubijciatur bafis
præclufa GHKL,
habens & ipfa in-
flexum fiphonem
M N X. & ofculo
F fubijciatur O P
infûdibulû, cuius
caulis feratur in
bafim G H K L,
tantum diftans à
fundo, quantum

Fig. 8.

ad

being that this is a more indulgent owl, or perhaps he is a teacher of singing, for in this case
the birds sing while he is looking at them and
cease the moment he turns his back.

Another pretty conceit of Hero's is shown in Fig. 8, in which there is a bird which not only makes a noise but at certain times will drink any liquid which is presented to it. The flow of water being intermittent, the cistern forming the pedestal is alternately filled and emptied. While it is being filled the air escapes through a whistle and causes the bird to sing, and when it is being emptied, by means of a siphon, a partial vacuum is produced and liquid presented to it is drawn up through the beak.

The next automaton from Hero is very ingenious and interesting, because it combines hydraulic, pneumatic, and mechanical actions. Here (Fig. 9) is a figure of Hercules armed with a bow and arrow ; there is also a dragon under an apple tree, from which an apple has fallen to the ground. Upon the apple being lifted, Hercules discharges the arrow at the dragon, which begins to hiss and continues to do so for some minutes. In this apparatus there is a double tank having a connection by a valve (H), which is attached

by a cord to the apple (κ), another cord, passing over a pulley, connects the apple with a

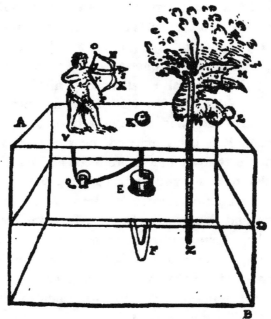

Fig. 9.

trigger in the right hand of Hercules. Upon
lifting the apple the trigger is released, and at

the same time the valve is opened, allowing the
water in the upper tank to flow into the lower,
by which means air is forced through a tube (z)
into the dragon's mouth, producing a hissing

Fig. 10.

sound, and this will continue until the upper
tank is empty. Here (Fig. 10) is Bate's version
of the same device, but very inferior to that from
which it was taken.

The next photograph is taken from another

work of Hero's, "*Quatro theorems aggiunti a gli artifitiosi spiriti,*" a copy of which I have here (Fig. 11), and which was printed at Ferrara in 1589.

This figure illustrates a very elaborate automaton, representing one of Vulcan's workshops in which you will see a smith forging a piece of iron, and assisted by three hammermen. The smith first puts his iron in the fire and then lays it on the anvil when the hammermen begin to hammer it ; then they leave off, and the smith turns round again to the fire. All these effects are produced by the machinery below the floor, and shown in the illustration. A shaft (A B) is driven by means of a water-wheel on the right, and on this shaft are projections or cambs which, by striking the ends of three levers (T, x, and v), pull the chains by which the arms of the hammermen are lifted. While this is going on the bucket (marked 20) is slowly filling, and when a sufficient weight of water has accumulated in it, it lifts the counterweight (17),

Fig. 11.

and, in doing so, rotates the vertical shaft to which the figure of the smith is attached, turning him round to the fire, and at the same time, by swinging round the conduit pipe (H I), cuts off the water from the wheel, and the hammermen cease to work until the smith is again ready for them. I think you will agree with me that this machine offers very fair evidence of the mechanical ingenuity of a man who flourished more than 2,000 years ago.

The last automaton of Hero to which I shall refer is perhaps the most ingenious of all, and it is one that those who were present when Brother Manning gave us his discourse on Robert-Houdin have already seen, I mean the little figure whose head cannot be severed from his body no matter how many times a knife be passed through his neck. Thanks to the kindness of my good friend I can show you one of these beautiful figures presented to me by him, and it will, I think, be of interest to him and to you to know that this device was invented nearly

2,000 years before Robert-Houdin was born, and a description of it with accompanying figures may be seen to-day in the British Museum in a Greek manuscript of the fifteenth century, which is a copy of Hero's Σπειραλια, and I now throw on the screen a carefully made facsimile (Fig. 12) of the figure given in that manuscript (which is known as No. 5605 of the Harleian Collection).

The head of this figure, which is otherwise separate from it, is attached to it by a peculiar shaped wheel pivotted between the shoulders of the body. This wheel may be described as a circular disc having an expanded rim so that a section taken through a radius would be of the form of the letter T, out of this wheel three nearly semicircular gaps are cut, each occupying sixty degrees of the circumference, and therefore leaving three portions of the rim, each also of sixty degrees. The neck attached to the head is fitted with a hollow T shaped circular groove into which the T ended arms of the wheel pass in succession as the wheel is rotated. As the groove in the head occupies nearly sixty

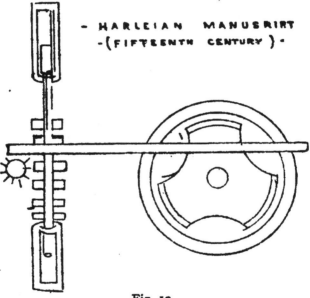

Fig. 12.

degrees it follows that as the wheel is rotated the rim of one arm can never leave the groove before the rim of the following arm has entered it, and so the head is attached to the body in every position of the wheel. When the knife is passed between the head and the body it strikes against one of the spokes of the wheel, moving it forward and pushing one of the arms out of the groove in the head, while, at the same time, another, following behind the knife, takes its place, and thus the head can never be detached from the body. Such an automaton is the little negro which I hold in my hand, for which I am indebted to the fraternal generosity of Brother Manning. Hero's description, however, carries the ingenuity of the device considerably farther, for in his automaton, not only is it impossible to sever the head from the body by passing a knife through the neck, but the figure can actually drink both before and after the operation. The illustration on the screen (Fig. 13) is a sort of modern restoration of the Harley drawing, showing the disposition of the various parts of the mechanism. A represents the wheel by which the head is held on to the body, and it will be

noticed that a tube D D leads from the mouth to
the neck and another, E, from the neck through
the body; these two tubes, marked respectively

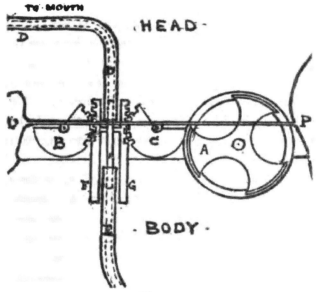

Fig. 13.

D D and E, are connected by the sliding tube F,
which is attached to the two racks F and G, into
which are geared the two toothed wheels B and

c. When the knife is passed from P to O it first rotates the holding-on wheel A, and then strikes against the radial face of the wheel C, turning it through a small arc, thereby moving the racks, and, sliding the connecting tube F out of D, allowing the knife to pass, which next strikes the radial face of the wheel B, and, by turning it, restores the sliding connecting tube F into D, and thus recompletes the connection. The sucking-up the liquid being accomplished in a similar manner to that in the drinking bird already described.[1]

I have now done with Hero of Alexandria, but, before passing to another period, I cannot resist showing you an invention of his which although not an automaton is too interesting in the light of modern civilization to omit. This (Fig. 14) is Hero's automatic penny-in-the-slot machine for giving a drink in exchange for a coin. If a "coin of five drachmas" be dropped into the slot it falls on a little plate at the end

[1] See page 30.

of a lever thereby opening a valve and allowing the liquid to escape through the nozzle.

It is more than probable that Hero was not himself the inventor of all the devices he de-

Fig. 14.

scribes, it is possible that many are due to Ctesibius whose pupil he was, and it is clear, from his own writings, that he was acquainted with the writings of Philo and of Archimedes. He was, however, the first to *describe* these inventions, and therefore it is only fair, in the absence of other evidence, to give him the credit.

There can be no doubt that puppets or dolls are of great antiquity; they were common with the ancient Egyptians, and here (Fig. 15) is an illustration of a doll from Thebes which is now in the British Museum, and you will notice that the head is covered with holes which served for the insertion of strings of beads to represent hair. Puppets were also in use with the Greeks, and afterwards found their way to Rome, and it is an interesting fact that, about three years ago, while the ground was being excavated for the foundations of the new Palais de Justice at Rome, at a spot not far from the Vatican, a stone coffin was discovered containing the skeleton of a young girl of about fifteen years of age, who had teeth of great beauty, and in her arms was a beautifully modelled wooden doll with jointed limbs which

Fig. 15.

was dressed in a rich material. The interment
had taken place in the time of Pliny, who refers
to the child, and mentions that she was engaged
to be married, a statement which is supported
by the fact that on one of the fingers is a doubly-
linked gold ring, besides other ornaments. The
coffin, with its contents as they were found, is
now in the museum in the Capitol and it is, I
believe, the only instance of an ancient doll
having been found in Rome, although moving
puppets or marionettes were known in very
ancient times, and are referred to by Xenophon,
Aristotle, Horace, Antoninus, Galen, and Aulus
Gellius.

The next figure is an illustration of what I
suppose must be the very earliest moving doll
in existence to-day; it is now in the Museum
van Oudheden at Leyden, and is a toy which
belonged to a child of ancient Egypt; I have
constructed a model of it by which you will
see that it is worked by pulling a thread; and
here I must make a passing reference to the

notorious phallic figures which were carried in procession during the festivals of Osiris and in the Dionysia of Bacchus. We are told by Lucian [1] that "Among the several sorts of Phalloi which the Greeks set up in honour of Bacchus there were figures of dwarfs with

Fig. 16.

moving parts actuated by strings, which were called 'Νευροσπαστα.'" In so eminently proper a community as we are in Ye Sette of Odd Volumes, I am unable to describe these figures in detail, or to exhibit them in action, but those who are *curious* as well as *odd* will find

[1] "De Syria Dea."

abundant evidence of them in the writings of Herodotus, of Lucian, of Pausanias, of Athenæus, of Plutarch, of Gyraldus, and of several other writers.

The earliest forms of moving puppets were set in motion by strings pulled by hand which were afterwards supplanted by cylinders turned by a winch, and the transition from that arrangement to the use of weights and springs was inevitable and was only a question of time.

From the time of Hero I have found nothing worth recording for nearly a thousand years, until the time of Charlemagne, to which monarch was presented by the Kalif Haroun al Raschid a most elaborate water clock. In front of the dial, and corresponding to the hours, were twelve little doors, and the time was shown by these doors opening one after another, each releasing a little brass ball which fell upon a small bell; after all the hours had struck, that is, at noon, another door opened, twelve little knights rode out, and, after career-

ing round the dial, they closed the doors and
retired. The eminent mechanician Gerbert who
occupied the papal chair in A.D. 1000, reign-
ing under the name of Silvester II., is said to
have constructed a speaking head of brass, and
was in consequence arrested for practising
magic, and Albertus Magnus, who flourished in
the thirteenth century, spent, according to his
own account, thirty years in the construction
of an automaton of clay which not only spoke
but walked and answered questions and solved
problems submitted to it. It is recorded that
his pupil, the celebrated St. Thomas Aquinas
was so horrified when he saw and heard this
figure that (believing it to be the work of his
Satanic Majesty), he broke it into pieces, when
Albertus cried aloud : "Sic periit opus triginta
annorum." I deeply regret this mischievous act
of St. Thomas Aquinas, because it renders it
impossible for me to show it to the Brethren
and our guests this evening. Roger Bacon also
is said to have made a similar automaton.

Records of speaking androides or talking
heads reach us from very early times. At
Lesbos there was a head of Orpheus which delivered oracles and predicted to Cyrus his violent
death, and we have it on the authority of Philostratus that the head was so celebrated for its
oracular utterances, among both the Greeks and
the Persians that even Apollo became jealous
of its fame.

Then again the mighty Odin had among his
mystical possessions a speaking head, believed
to be that of Minos, which Odin preserved by
encasing it in solid gold. He is said to have
consulted it on all occasions, and its utterances
were regarded as oracles.

Mention might here be made of the colossal
figure of Amunoph III. on the plain of Thebes,
and which is commonly known as the "vocal
Memnon," of which a photograph is now before
you; it is the more eastern of the two Colossi,
and, when the first rays of the morning sun fell
on it, it emitted a sound which has been de-

scribed as similar to that of the snapping of a harp string, but it has been silent since the time of Severus. It is a seated figure nearly sixty feet in height, and is in no sense an automaton, but I mention it here because it was believed to utter sentences which the ancient priests of Egypt alone, for the very best of reasons, knew how to interpret.

In more modern times we hear of the eminent Dr. Wilkins, Bishop of Chester (who married the sister of Oliver Cromwell, and who may be regarded as the founder of the Royal Society), experimenting upon the transmission of sound; and Evelyn, in his "Diary," writing on the 13th of July, 1654, says, "We all dined at that most obliging and universally curious Dr. Wilkins's, at Wadham College. He had contrived a hollow statue, which gave a voice and uttered words"; and in his "Mathematicall Magick," (a copy of which I have here) which was published in 1648, Wilkins refers to the speaking figures of the ancients.

A contemporary of Wilkins was the celebrated Edward Somerset, Marquis of Worcester, who in his "Century of Inventions" gives as his 88th device: "How to make a Brazen or Stone-head in the midst of a great Field or Garden, so artificial and natural that though a man speak never so softly, and even whispers into the eare thereof, it will presently open its mouth, and resolve the Question in French, Latine, Welsh, Irish or English, in good terms uttering it out of his mouth, and then shut it untill the next Question be asked."—But, unhappily, he does not tell us how it may be done.

The great period for the construction of automata began at the close of the fourteenth century, and reached its climax at the end of the seventeenth and beginning of the eighteenth century. One of the earliest mechanicians who devoted his skill to automata was Johann Müller, of Königsberg, commonly known as Regiomontanus. This eminent mathematician and astronomer made of iron a fly which is said to

D

have left his hand and, after flying to each of the guests in the room, returned to its master, alighting on his hand. Müller made also a still more wonderful machine; this was an artificial eagle which, on the authority of Peter Ramus, flew to meet the Emperor Maximilian on his entry into Nuremberg on the 7th of June, 1470. After soaring aloft in the air, Ramus informs us, the eagle met the emperor at some distance from the city, then returned and perched upon the city gate where it awaited the emperor's approach. On his arrival the bird stretched out its wings and saluted him by bowing.

It is a remarkable fact that not one of Müller's contemporaries, who often refer to this learned man and to his great accomplishments, makes any reference to these pieces of mechanism, and Peter Ramus was not born until forty-five years after, but they are referred to by Baptista Porta, Gassendi, Lana, and Bishop Wilkins, who, however, differ considerably in their dates. Strada, in his "De Bello Belgico," tells us that the

Emperor Charles V., after his abdication in 1556, took a most keen interest in automata of various kinds, and he employed a very skilful artist, Janellus Turrianus, of Cremona, to construct them for him. This mechanic made figures of horsemen which marched along the table, played upon flutes and drums, and entered into combat with one another, and he exhibited wooden birds which flew up to their nests (they must, I think, have been *wood pigeons*). This Janellus Turrianus was evidently a very wonderful man, for he made a corn-mill so small that it could be concealed in a glove, and yet could grind in a day as much corn as would supply eight men with food. I never saw this machine myself, and I cannot help thinking that either the glove must have been rather large or the appetites of the men must have been rather small. Apart, however, from the exaggeration of the genius of this man, he was undoubtedly a most skilful mechanician, for he repaired and considerably improved a

very curious clock, the work of Martinot, a
clockmaker of the seventeenth century.   Before it struck the hour two cocks flapped their
wings and crowed alternately, then two little
doors opened and a figure came out of each
carrying a gong which was struck by armed
guards with their clubs.   These figures having
retired, a door in the centre opened and an
equestrian figure of Louis XIV. came out.   At
the same time a group of clouds separated
giving passage to the figure of Fame which
hovered over the head of the king.   An air was
then chimed upon the bells, after which the
figures retired; the two guards raised their
clubs and the hour was struck.   . .

   In the year 1788, Agostino Ramelli published
his important work " *Le diverse ed artificiose
Machine,*" and I have reproduced some of the
plates in that beautiful book, a copy of which is
before me. (one of which, Fig. 17, see *Frontis-*
piece, I have chosen to adorn the menu which is
on the table, for no other reason than that it

Fig. 18.

appeared especially appropriate as figurative of
the desire of your humble Mechanick to be for
ever associated with Ye Sette of Odd Volumes).

In the next illustration (Fig. 18) we have a
beautiful plate from Ramelli, in which another
of Hero's inventions, the group of singing birds
is introduced as an ornament in an elaborately
furnished room of the period. In this case the
water is in the first instance lifted by air being
blown in through a pipe by a person concealed
behind the wall which in the drawing is broken
away to show a mediæval old buffer engaged in
this manly performance.

About the middle of the seventeenth century
magnetism began to be employed for producing
the effects of magic, and that extraordinary ver-
satile all-round Odd Volume, Athanasius Kircher, in his "Magnes sive de Arte Magnetica,"
which was published in 1641 (a copy of which
I have here), describes and illustrates several
automata which depend for their action upon
magnetism. Here, for example (Fig. 19), he

gives a representation of the Dove of Archytas, which by the action of a revolving loadstone, is made to fly around a dial and mark the hours by pointing to the figures on its edge.

Time will not permit me to say as much

Fig. 20.

about this curious old book as its quaintness and terribly bad science deserve, I will only show you one more illustration from it in which a wheel is driven round by two Æolipiles in the form of human heads, which blow out jets of steam against the cellular periphery of the wheel, and in the lower figure the little boilers (c and

Fig. 19.

D) which the heads inclose, are shown separately, the nozzle of one pointing upwards, while that of the other has a downward direction.

When Kircher's book was published Louis XIV. was a child, and it is stated by several authorities that both Père Truchet and Camus made the most elaborate automata for his boyish amusement, but as Louis XIV. was forty years old when Truchet came of age and fifty-five when Camus was twenty-one it is difficult to reconcile these statements with facts.

Putting aside, however, the question of the period of life when the king amused himself with such things, it is well authenticated that Père Truchet, towards the end of the seventeenth century, constructed for him moving pictures which exhibited extraordinary mechanical skill. One of these was the representation of a five-act opera, the scenery of which was automatically changed between the acts. The actors came on and went off, and performed their parts in pantomime. The proscenium

was about sixteen inches in breadth and thirteen
in height, and the whole of the machinery with
the scenery occupied a space only an inch and
a quarter in depth.[1]

The account given by Camus of a toy he
constructed for this baby king of fifty summers
is very wonderful. This elaborate automaton
consisted of a small coach drawn by two horses
and which contained the figure of a lady with a
footman and a page behind. When this little
coach was placed on the edge of a suitable table
the coachman smacked his whip and the horses
immediately started, moving their legs in a most
natural manner; when they reached the opposite edge of the table they turned sharply at
right angles and proceeded along that edge. As
soon as the carriage arrived opposite the king
it stopped and both the footman and page got
down and opened the door, the lady alighted,
and, curtseying to the king, presented a peti-
tion. After waiting a few minutes she bowed

[1] Mem. Acad. Sc. Paris, 1729.

again to the king and re-entered the carriage, the page got up again behind, the coachman whipped up his horses and drove on, and the footman running after the carriage jumped up into his former place. In the account given by M. de Camus he does not attempt to describe the mechanism of the machine and we have his word alone for the account of its performance.

The great philosopher Descartes formed the theory that all animals are merely automata of a high degree of perfection, and, to prove his notion, he is said to have constructed an automaton in the form of a young girl to which he gave the name of " Ma fille Francine." This figure came unhappily to a watery grave, for during a voyage by sea the captain of the vessel in which it was travelling had the curiosity to open the case in which Francine was packed and, in his astonishment at the movements of the automaton, which were so wonderfully natural, he threw the whole thing overboard, believing it to be the work of the devil.

I now come to what are, if not the most extraordinary *pieces of mechanism*, certainly the most wonderful *automata* the world has ever seen. In the year 1738 that great mechanical genius M. Vaucanson, a member of the Académie des Sciences exhibited at Paris three very remarkable automata which were, a flute-player, a figure which played the shepherd's pipe of Provence and the drum, and an artificial duck. The first of these, the flute-player, he described in a Memoir read before the Académie on the 30th of April, 1738. This automaton was a wooden figure six feet six inches in height, repre-senting a well-known antique statue of a Faun, sitting on a rock and mounted on a square pedestal four feet six from the ground. It was capable of performing twelve pieces of music on a German flute, the instrument being really played as a man would play it by blowing across the embouchure and projecting the air with variable force by movable lips, which imitated in their action those of a living player, employing a tongue to regulate the opening, and producing the notes by the tips of the fingers closing or opening the holes.

The mechanical devices in this automaton are so beautiful and so scientifically thought out, that I am only sorry that time will not permit me to describe them in detail, but I will try and make its general principles clear.

Within the pedestal was a train of wheel-work driven by a weight, which set into motion a small shaft on which were six cranks disposed at equal angular distances around it; to these six cranks as many pairs of bellows were attached (their inlet valves being mechanically opened and closed so as to make them silent in action). The air supplied by these bellows was conveyed to three different wind chests, one loaded with a weight of four pounds, one with a weight of two pounds, and the last having only the weight of its upper board. These wind chests communicated with three little chambers in the body of the figure, and these chambers were all connected with the windpipe which passed up the throat to the cavity of the mouth and terminated in the two movable lips which, between them, formed an orifice that could be protruded or drawn back, and might be further modified by the action of the tongue.

The train of wheels also set into motion a cylinder twenty inches in diameter and two feet six inches long; on this were fixed a number of brass bars of different lengths and thicknesses which in their revolution acted upon a row of fifteen keys or levers; three of these corresponded to the three little wind chambers containing air at different pressures, and, by means of little chains, operated their respective valves. There were seven levers set apart for operating the fingers, their respective chains making bends at the shoulders and elbows of the automaton, and terminated at the wrist in the ends of what I may call metacarpal levers attached to the fingers which were armed at their tips with leather to imitate the flesh of the natural hand.

The motion of the mouth was controlled by four of the levers, one to open the lips so as to give to the wind a greater issue, one to bring them closer together, and so contract the passage, a third to draw the lips backward and away from the flute, and the fourth to push them forward over the edge of the embouchure.

The last of the fifteen levers is the cleverest of all, for it has the power of controlling the tongue, an accomplishment which I think everyone (unless he be an Odd Volume) will agree with me is a very difficult one to acquire.

The barrel worked upon a screwed bearing (similar to that of the cylinder of a phonograph), so that in its revolution all the levers described a spiral line sixty-four inches long, and, as the barrel during the performance made twelve revolutions it followed that the levers passed over a distance of no less than 768 inches in going through its performance of twelve tunes.

In a Memoir read before the Académie des Sciences, M. Vaucanson described the very

beautiful methods by which the barrel was set
out, and by which the positions of the bars were
determined on its surface so as to regulate the
supply of air and to control the actions of the
fingers, the motion of the lips and the movements of the tongue; and he gave a most
interesting analysis of the acoustics of wind instruments; but time will not permit me to make
more than this passing reference to them.

The picture on the screen (Fig. 21) is a photographic reproduction of the plate attached to
M. Vaucanson's Memoir (a somewhat rare little
tract published in 1738) in which his three
automata are shown, and I hold in my hand a
copy of the translation by Dr. Desaguliers,
published in London in 1742, which, the im-
print tells us, was "*sold at the long room at the
Opera House in the Haymarket, where the mechanical figures are to be seen at 1, 2, 5, and 7
o'clock in the afternoon.*"

The second of Vaucanson's automata was his
celebrated model of a duck, which he himself

Fig. 21.

described in a letter to the Abbé de la Fontaine in 1738. This extraordinary automaton (according to the inventor's own account of it), exhibited a considerable amount of physiological and anatomical knowledge and the most profound mechanical skill, for in it the operation of eating, drinking, and digestion, were very closely imitated. The duck stretched out its neck to take corn from the hand, it swallowed it and discharged it in a digested condition, the digestion being effected not by trituration, but by dissolution, and (to quote the quaint expressions of the inventor), "The matter digested in the stomach is conducted by pipes (as in an animal by the guts), quite to the anus, where there is a sphincter that lets it out. I don't pretend," he says, "to give this as a perfect *digestion*, capable of producing blood and nutritive particles for the support of the animal. I hope nobody will be so unkind as to upbraid me with pretending to any such thing. I only pretend to imitate the mechanism of their action in

E

these things, *i.e.*, first, to swallow the corn; secondly, macerate or dissolve it; thirdly, to make it come out sensibly changed from what it was." But (on the same authority), besides being furnished with a digestive system, the wings were anatomical imitations of nature; not only was every bone imitated, but all the processes and eminences of each bone, and the joints were articulated as in a real animal.

After having been wound up, the duck ate and drank, played in the water with his bill, making what is described as a "gugling" sound, rose up on its legs and sat down, flapped its wings, dressed its feathers with its bill, and performed all these different operations without requiring to be touched again.

It is important, however, to point out that this digestion story can only be "digested" *cum grano salis*, and this is supplied in the sequel which furnishes the explanation. In the year 1840 the automaton was found hidden away in a garret in Berlin; it was very much out of

order, and a mechanician of the name of
Georges Tiets undertook to repair it. It was
taken to Paris, and in the year 1844 was exhibited in the Place du Palais Royal. In the
course of this exhibition one of the wings
became deranged, and it was put into the
hands of Robert-Houdin for repairs. Robert-
Houdin took advantage of this opportunity for
examining the so-called digestive system of the
automaton, and he thus describes its action:

"On présentait à l'animal un vase dans lequel
était de la graine baignant dans l'eau. Le
mouvement que faisait le bec en barbotant
divisait la nourriture et en facilitait l'introduc-
tion dans un tuyau placé sous le bec inférieur
du canard; l'eau et la graine, ainsi aspirés tom-
baient dans une boîte placée sous le ventre de
l'automate, laquelle se vidait toutes les trois ou
quatre séances. L'évacuation était chose pré-
parée à l'avance; une espèce de boullie, com-
posée de mie de pain colorée de vert, était
poussée par un coup de pompe et soigneuse-

ment reçue, sur un plateau en argent, comme produit d'une digestion artificielle," so that, after all, this wonderful digestion of Vaucanson's duck was nothing more than a clever trick.

The third automaton of Vaucanson was a figure that played on a shepherd's pipe with one hand while it beat a drum with the other. The instrument played upon was a little pipe with only three holes, and the different notes were produced by a greater or less pressure of air and a more or less closing of the holes, and every note, no matter how rapid was the succession, had to be modified by the tongue. In this machine there were provided as many different pressures of air as there were notes to be sounded, and the mechanism by which these operations and the fingering of the keys were effected reflects the greatest credit on the memory of this remarkable man.[1]

[1] Beckmann in his " History of Inventions," says that these automata found their way to St. Petersburg, and that in 1764, he himself saw them at the Palace of

The Automaton duck of Vaucanson was, to a certain extent, anticipated by the Comte de Gennes, Governor of the Island of Saint Christopher, who, we are told by Père Labat, constructed a peacock which could walk about and pick up grains of corn, which it swallowed and digested. I have no means of determining whether or not Vaucanson took the idea of his duck from this automaton, but that Vaucanson had imitators there is abundant evidence to prove. In the year 1752, Du Moulin, a silversmith, travelled all over Europe with automata similar to those of Vaucanson, and they were afterwards purchased in Nuremberg, by Bereis, a counsellor of Helmstädt, at whose place they were seen by Beckmann in 1754.

In the year 1760, there was a writing automaton exhibited in Vienna, which was constructed by Friedrich von Knaus, and about the

Zarsko-Selo, where he learnt that they had been purchased from Vaucanson, but they were not, at that time, in working order.

same time a number of very curious automata were made by Le Droz, of Chaux de Fonds, in Neufchatel. One of these was a clock, presented to the King of Spain, which had, in addition to several moving figures, a sheep that bleated in a very natural way, and a dog mounting guard over a basket of fruit ; if anyone attempted to touch the basket the dog barked and growled, and if any of the fruit were taken away the barking continued until it was restored.

The son of this man (who lived at Geneva), was no less skilful a mechanician, for he made a gold snuffbox about $4\frac{1}{2}$ inches long by 3 inches broad, in which when a spring was touched a little door flew open and a beautifully modelled bird of green enamelled gold rose up, fluttered its wings and tail, and commenced a trilling song of great beauty and power, its beak keeping time with the notes. Such a snuffbox was exhibited in the Great Exhibition of 1851, proving as great a popular attraction as the Koh-i-nur diamond, and (owing to the kindness of my friend Mr.

Tripplin the well-known horologist) I am now able to show you one of these very beautiful triumphs of mechanical skill.

Another of the younger Le Droz's inventions was his celebrated drawing automaton, which was a life-size figure of a man sitting behind a table and holding a style in his hand. A sheet of vellum was placed on the table, and the figure began to draw portraits of well-known persons with extraordinary correctness. This automaton was shown in London, and attracted considerable attention at the time.

I must now re-introduce to you another old friend, first shown here by Brother Manning. Here he is! a little acrobat that turns somersaults backwards down stairs. This is not, as many have thought, an invention of that great mechanical genius, Robert-Houdin, for it is figured and described in Musschenbroeck's "Introductio ad philosophiam naturalem," which was published in Leyden in 1762 (a year after the author's death), and half a century before

Robert-Houdin was born, and on the screen you have a facsimile (Fig. 22) of Musschen-broeck's illustration of this mechanical toy, which he refers to as "an old invention of the Chinese." It is also described by Ozanam in his "Recréations Mathématiques et Physiques,"

Fig. 22.

the first edition of which was published in 1694. The figure I now throw on the screen (Fig. 23), is taken from the second edition of this work which was edited by Montucla in 1790. The principle is exceedingly simple; the whole thing depends upon the centre of gravity being suddenly changed by a shifting weight. Within a

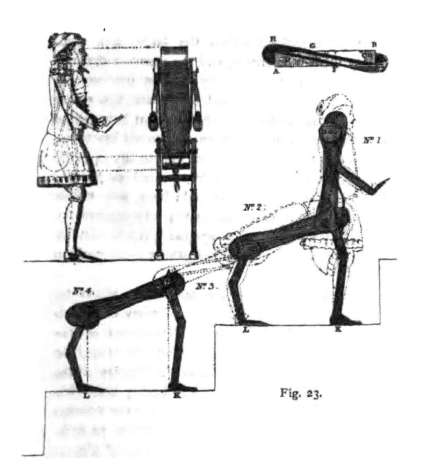

Fig. 23.

tube contained within the body, is a small quantity of mercury, and the moment that this tube is inclined to the horizon the mercury flows to the lower end tilting one figure over the other, and with such force that it is carried over by its inertia far enough to tilt the tubes, and cause the mercury to flow to the opposite end, and the process is repeated as long as there are stairs to descend; by a very simple arrangement of strings passing over pulleys, the legs and arms are always brought into suitable positions to support the figure in every position of its descent.

I now come to the automaton which for some years was the wonder of every country in Europe, the automaton chess-player of the Baron Wolfgang von Kempelen, constructed in 1776. This automaton was a life-size sitting figure dressed as a Turk, and having before it a large rectangular chest or cabinet, 3 feet 6 inches long, 2 feet deep, and 2 feet 6 inches high, on the top of which was a chessboard and a set of men. The seat on which the figure sat, was attached to the cabinet and the whole was on castors, so that it could be wheeled about the floor. When the automaton was exhibited, the exhibitor began operations by opening the doors of the cabinet so as to show its contents, and here I will throw on the screen a copy (Fig. 24) of one of the plates in a curious pamphlet,[1] printed anonymously in 1821, but probably by Professor Willis. It must, however, be recollected that these doors were opened in succession, and never all at the same time, but whichever door was opened, nothing could be seen but wheels, levers, connecting rods, strings and cylinders. After this the doors were closed and locked, the machinery was wound up, and the figure was ready to play a game of chess with any one who would challenge him. On

---

[1] "An Attempt to Analyse the Automaton Chess Player of Mr. de Kempelen, with an easy method of imitating the movements of that celebrated figure. Illustrated by original drawings. 8vo. London. 1821."

commencing the game the figure moved its head, and seemed to look at every part of the board. When it checked the king, it nodded its head three times, and when it threatened the queen, it nodded twice. It also shook its head when its adversary made a false move, and replaced the offending piece. It nearly always won the game, but occasionally lost.

When it was completed, it was exhibited in Riga, Moscow, St. Petersburg, Berlin, Presburg and Vienna, coming to London in 1783, and having been seen by many thousands during those years with out its secret being discovered, but in the year 1789, a book was published by Mr. Freyherre of Dresden, in which he showed that "a well taught boy very thin and tall of his age, (sufficiently so that he could be concealed in a drawer below the chess-board,) agitated the whole." In the plate before you, you will see that the author has shown in dotted lines, the position a boy might take when the left hand door was opened.

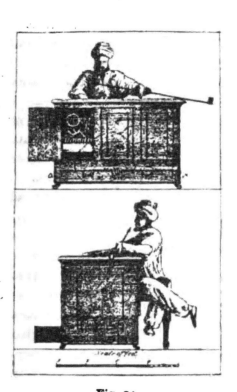

Fig. 24.

The real story of this most ingenious and suc-
cessful scientific fraud is so interesting that
I must tell it here, although it puts for ever
Baron von Kempelen's chess-player outside the
circle of true automata.   In the year 1776, a
regiment, half Russian and half Polish, mutinied
at Riga.   The mutineers were defeated,' and
their chief officer, Worouski, fell, having had both
his thighs fractured by a cannon ball.   He hid
himself in a ditch until after dark, when he
dragged himself to the neighbouring house of a
doctor named Osloff, a man of great benevo-
lence, who took him in and concealed him, but
he had to amputate both his legs.   During the
time of Worouski's illness, Osloff was visited by
his intimate friend the Baron von Kempelen,
and after many consultations and much thought,
Kempelen hit upon the idea of conveying him
out of the country by devising this automaton
(as Worouski was a great chess-player), and in
three months the figure was finished.

In order to avoid suspicion he gave performances *en route* to the frontier. The first performance was given at Toula, on the 6th of November, 1777 (that is to say exactly 114 years ago to-day). The machine and Worouski were packed in a case and started for Prussia, but when they reached Riga, orders came from the Empress Katherine II., for Baron von Kempelen to go to St. Petersburg with his automaton. The Empress played several games with him, but was always beaten, and then she wanted to buy the figure. This was an awkward situation for Kempelen, and he was at his wits' end to know how to wriggle out of it. He declared that his own presence was absolutely necessary for the working of the machine, and that it was quite impossible for him to sell it, and, after some further discussion, he was allowed to proceed on his journey.

This chess-player was, in the same year, purchased by Mons. Anthon, who took it all over Europe. At his death it came into the hands of Johann Maelzel, the inventor of the Metronome, who sent it to the United States. It was afterwards sent back to Europe, and in the year 1844 was in the possession of a mechanician of Belleville, named Croizier.

Maelzel himself was a mechanician of very considerable skill, and he constructed an automaton trumpeter, which was exhibited at Vienna about the year 1804, which played the Austrian and French cavalry marches, and marches and allegros by Weigl, Dussek, and Pleyel. Maelzel was, after that, appointed mechanician to the Austrian Court, and constructed an automatic orchestra, in which trumpets, flutes, clarionets, violins, violoncellos, drums, cymbals, and a triangle, were introduced, and this attracted very great interest in the Austrian capital at the time.

In the year 1772 there was in Spring Gardens, near Charing Cross, a most remarkable collection of automata exhibited in a place of entertainment known as Cox's Museum, and here I have an original copy of the "*Descriptive catalogue, of the several superb and magnificent pieces of mechanism and jewellery exhibited in Mr. Cox's Museum, at Spring Gardens, Charing Cross."* To which this footnote is added, *" Hours of Admission,* 11, 2, *and* 7, *every day (Sundays excepted), tickets Half a Guinea each, admitting one person, to be had at Mr. Cox's, No.* 103, *Shoe Lane."* This was a very extraordinary exhibition, and contained upwards of twenty large and elaborate automata, several of them being adorned with gold and precious stones. Some were complicated clocks, some were large groups of animals, and figures with fountains and cascades around them. None of these objects was less than nine feet high, and some were as high as sixteen feet. I can find nothing important enough from a Mechanick's point of view, to describe in detail, but it was the precursor in the same place of the exhibition of Monsieur Maillardet, which was one of the London attractions at the beginning of the present century.

F

M. Maillardet exhibited a bird automaton
(similar to that already referred to which was
made by Le Droz), and whose performance
lasted four minutes with one winding up. He
constructed also a spider, entirely of steel,
which imitated all the actions of the real animal,
it ran round and round the table in a spiral line,
tending towards the centre. Maillardet made
automata representing a caterpillar, a mouse, a
lizard, and a serpent; the last crawled about all
over the table, darted its tongue in and out, and
produced a hissing sound.

Maillardet's most important automata were,
however, his drawing and writing figure, and his
pianoforte player. The former was a kneeling
boy, who wrote in ink with an ordinary pen,
sentences in English and in French, and drew
landscapes. The pianist was a figure of a lady,
who performed eighteen pieces of music. She
began by bowing to the audience, her bosom
heaved, and her eyes first looked at the music,
and then followed the motion of her fingers, and

the music was produced by the keys being played
on by the fingers; but the most remarkable
of M. Maillardet's machines, was a magician, or
fortune-teller, which gave answers to some
twenty given questions, which were inscribed on
as many counters or medallions. One of these
medallions having been put into a drawer, the
figure arose from his seat, bowed to the audience,
and described mystic circles in the air with his
wand; after appearing to consult his book of
mysteries, he struck a little door behind him,
which flew open, and exhibited an appropriate
answer to the question on the medallion.

The general principle upon which this automaton's power of selection was founded lay in
the fact that in the edge of each medallion there
was a small hole drilled, but no two holes were
drilled to the same depth, and, by an exceedingly delicate mechanism, the varying depth to
which a pin could be thrust into the edge of
a disc, was caused to control the mechanism by
which the various answers were selected, and

which were exhibited when the little door flew open.

The next great master of automaton design and construction, was that wonderful genius Robert-Houdin (about whom our worthy Secretary and Seer discoursed to us so pleasantly and so instructively nearly a year ago). Brother Manning's paper was so complete in itself, and that part of it which dealt with automata was so ably illustrated, that it will be quite unnecessary for me to add to the length of this communication, by going over that ground again, so I will merely enumerate the automata of that interesting man and pass on to still more recent times.

The first of the automata of Robert-Houdin was a confectioner's shop, in which a pastry-cook came out of the door when requested and offered to the spectators patisserie, bonbons, and refreshments of every description, and within the shop might be seen the assistants making pastry, rolling out the dough, and putting it into the oven. Then he made two clowns, known as

Auriol and Débureau. The first of these performed a number of acrobatic feats upon a
chair which was held at arm's length by the
other. After this, the figure of Auriol smoked
a pipe, and accompanied on the flageolet an air
played by the orchestra.

Another was an acrobat which performed
tricks on the trapèze, and the last to which I
shall refer, was his celebrated writing figure,
which is illustrated in Brother Manning's "Opus-
culum," No. XXIV., to which I must refer you
for a great deal of interesting information re-
specting that remarkable man.

A contemporary of Robert-Houdin was Mons.
Mareppe, who constructed a very wonderful
automaton violin player, and which was exhibited at the Conservatoire at Paris, in the
year 1838, and which performed on the violin
by bowing and fingering the strings, and in
an account of the performance which was published at the time in "Galignani's Messenger,"
it is stated that the musical execution was so

perfect as to bring tears into the eyes of the audience.

Coming to our own period, from the time of Robert-Houdin, there have been no great automata which will live in the history of the subject, until the year 1875, when Mr. J. N. Maskelyne (who, I am happy to tell you, is honouring us with his presence to-night) exhibited at the Egyptian Hall his marvellous " Psycho." This was a seated figure, supported by a cylindrical pedestal of glass which stood upon a little platform, and, being on castors, could be wheeled about the floor. This automaton can actually play a game of whist, selecting the cards from a rack in front of it, and playing a most skilful game. The machine works apparently without any mechanical connection with anything outside, and the delicacy and precision of its actions, display the most consummate skill in design, and give to its inventor a high position for mechanical science. This automaton also works out arithmetical calculations, with numbers from one to a hundred millions, showing the result behind a door which opens in front of its box.

Another of Mr. Maskelyne's automata, is the celebrated "Zoe." of 1877, a sitting figure supported like the last on a glass pedestal so as to exclude the possibility of an electrical system of communication. A sheet of paper is fastened on to the table in front, and the figure traces out very fair portraits of public characters chosen by the audience out of a list of some two hundred names.

In respect to these most beautiful machines I must refrain from revealing to you the secrets of their working, and that for two reasons, first, because I do not know them myself; and second, because Mr. Maskelyne is here and is doubtless only impatient to jump up when I sit down and tell us all about them.

I do not intend to say anything about speaking machines or to do more than make a passing reference to the very interesting work and researches of Kircher in 1650, Van Helmont,

1667, Kratzenstein, in 1780, L'Abbé Mical, in 1783, Von Kempelen in 1791, Willis in 1829, Wheatstone in 1837, or of Faber in 1862. All these mechanicians and physicists studied the philosophy of speech and produced machines or parts of machines, which could utter vowels, words or even sentences, but these machines were operated by keys and stops and were, in no sense of the term, automata.

I must, however, refer to one of the greatest marvels of modern science, the phonograph which Mr. Edison has applied in the construction of his talking dolls. Edison's talking doll is a figure, within which a little phonograph, driven by a little winch, is concealed, and which repeats in a clear voice any sentence or rhyme which may have been spoken against its recording cylinder or disc. I am deeply disappointed to be unable to show you one of these most inte-resting automata to-night, for one is on its way to me across the Atlantic. Colonel Gourand very kindly sent for one that I might show it to

you this evening, and I deeply regret that it has
not arrived in time, for the Odd Volumes would,
otherwise, have been the first to hear its voice in
Europe.[1]

In the phonograph, that splendid triumph of
acoustical and mechanical science, we have
literally fulfilled, the prediction made by Sir
David Brewster in 1883, when he wrote " I
have no doubt that before another century is
completed, a talking and a singing Machine will
be numbered among the conquests of Science."

No one who is familiar with any of the great
European capitals can have failed to notice in
the windows of the higher class of toy-shops,
clock-work automata of various kinds. We have
jugglers and rope dancers, conjurers, pianists,
violinists, harpists and trumpeters, dancing nig-
gers, figures fighting, knitting, sewing, writing,
and engaged in almost every occupation per-

[1] The author exhibited Edison's talking doll at the
Conversazione of the Sette of Odd Volumes which was
held the following month.

formed by human beings, but none that I have
seen are fit for comparison with the wonderful
mechanical works of Vaucanson, Robert-Houdin
or Maskelyne ; mechanically they are nearly
identical with one another, and differ only in the
external application of the internal machinery.
At International Exhibitions one sees one or two
of superior merit, but I have not recently seen
any of sufficient importance to bring before you
this evening.   The pianists and other musicians
merely move their hands on their instruments,
but the music (save the mark) whether it be a
violin or a trumpet, comes from a musical snuffbox inside which is generally wound up by a
different key.   These figures are usually very
costly, and I am always puzzled to know who are
the people who purchase them.   The best are
generally those mechanical toys which represent
the movements of animals, and here I have a
mechanical bear which is rather amusing, and it
is ingenious because by a very simple combination of clock work with cranks and strings a

number of different motions is obtained; we
have the mouth opening and shutting, the head
going from side to side, the lips moving and the
whole animal bowing to the spectators.

Within the last few years a most extraordinary
amount of mechanical ingenuity has been brought
to bear upon the construction of small automatic
toys, which are sold in the streets for a few pence,
and I think, even more than the extraordinarily
simple and ingenious contrivances by which the
various effects are produced, the great inventive
merit consists in a design and method of manu-
facture by which they can be turned out, with
a profit, at so insignificant a cost. I have
brought together a few examples, a very minute
fraction of the hundreds of forms that exist, but
selected merely to illustrate the different types
of principle of action.

A very favourite motive power is a wound up
spring, consisting of strands of vulcanized india-
rubber, and here I have one of the well-known
butterflies which come out in Paris in 1878,

where they filled the air of the Avenue de l'Opera, the shops of which were then occupied chiefly by hawkers of toys. The motive power of this toy is nothing more than a light screw propeller or fan rotated by the untwisting of a spring, while on the body of the machine are two fixed wings or fins to prevent the whole machine from rotating. The action is wonderfully like that of an animal, perhaps most like that of a bat. Here again the same principle is applied in a running mouse, and this is especially interesting from the fact that the machine winds itself up the moment the tension of the cord is relaxed, and as the spindle of the wheels is the flexible rubber itself, the peculiar scuttling action of a mouse is well imitated.

There is again a large class of mechanical toys in which there is a combination of a rubber spring with a wheel and escapement, the pallets of which by their reciprocating motion producing, whatever effect may be desired; the swimming fish is one of them, the wagging of the tail being produced in the way I have described. Here is another displaying considerable ingenuity. In this case an escapement wheel works a crutch which by a pair of cranks linked together causes each of two pugilists to turn a little way backwards and forwards on one heel, and the arms being hung loosely to the shoulders by rubber hinges give to the figures the appearance of hitting out vigorously.

I have here a couple of figures which I admit do not contain their motive power within themselves but they require so little aid from outside and do so much for themselves that I have been tempted to bring them in. Here is a monkey climbing a rope, and its progression is insured by the simplest possible device, the string passes over one pin and under another in his posterior hands while the anterior pair of hands grip the rope with a slight degree of friction : if the string be tightened the lower hands act as a lever which pushes the body up, but when it is slack it slips round the pins and does no work, in other words the grip of the hands is greater than that of the feet when the cord is slack but less when it is tight.

In this little animated skeleton, we have an immense effect produced by an extraordinarily small external motion. The squeeze that I give to this U shaped spring, by varying the tension of the twisted strings, on which the skeleton is suspended, is almost infinitesimal—but it gives to the skeleton considerably more energy than is usually to be found in skeletons.

Here we have a walking figure whose action depends upon gravity, but his progression is checked by the friction of his feet on the board on which he performs, first one foot catches and then another, and each time his inertia turns him round, which gives him an appearance of having been in the company of teetotallers, or can he have been dining with the Sette of Odd Volumes?

A familiar form of mechanical or automatic toys is in the form of a box or frame having a glass front, behind which figures of acrobats, rope-dancers and moving groups are set into motion by sand falling on a wheel within the case; and it is an ingenious feature of these toys that they are "wound up" by simply rolling the box over on its edge through one revolution, which has the effect of lifting the fallen sand back into the upper reservoir.

The last great class of mechanical figures, to which I shall refer, includes those which depend for their action upon the spinning of a top or fly-wheel, and some of them are particularly pretty and ingenious.

Here, for example, is a couple of figures, which the gentleman who sold it to me told me was "a Narry and a Narriet walking hout on 'Ampstead 'Eath." In this case the ruling spirit and go is as usual in the *lady*, and the man has to follow whither she leads, the legs of the man are connected together at the hips by a pair of cranks so disposed, that if one leg be pushed back, the other is thereby thrown forward.

Now the heels are so cut that they catch in the
ground when in a forward position and can slide
forward when behind; in being urged along, the
forward leg catching in the ground is relatively
pushed back and the other leg comes forward,
which in its turn catches, and the effect of
walking is produced.

And here we have (Fig. 25) another on precisely the same principle, in which an ostrich
appears to draw a cart, which in reality, is
pushing him along, but the effect of the ostrich's
strut is delightfully reproduced.

Here is another in which several very curious
motions are reproduced. This beautiful little
mechanical toy (Fig. 26) represents a circus girl
riding round the ring, and occasionally leaping
over a bar or bowing to the audience, while the
prancing action of the horse is ingeniously imitated. The motive power is derived from the
spinning of a top or fly-wheel, supported in a
frame attached to the bar to which the horse is
fixed; and, as the spindle of the top spins on

Fig. 25.

the bevel edge of the circular base, the horse is caused to gallop round in a circle, and, being supported on the table by a roller mounted eccentrically on its axis, it prances up and down as it runs. The equestrienne is attached to a light lever pivotted on the rotating frame and revolving with it. Twice in its revolution this lever is lifted by a cam, forming part of the base; the first lift causes the figure to give a little bow, and the second, which is much greater, makes her leap over the bar under which the horse runs. This little machine is one of the most mechanically ingenious of the modern automaton toys, and it is made at the cost of only a few pence.

The last I shall show you is this elephant. In this little machine we have a fly-wheel, which with its vertical shaft looks like an umbrella over the Nabob who sits on the top, the vertical shaft passes into the body of the elephant, and there by a simple frictional gearing, rotates a couple of cranks to which the legs are connected. The effect of spinning the umbrella is therefore merely to move the legs backwards and forwards; and, if that were all, no progression could be effected; but each foot rests on a little wheel or roller, which can only rotate in one direction, so that while it catches the ground in its backward stroke it rolls freely over it while it is moving forward, and thus each leg in its turn contributes to the progressive movement of the toy.

Now I have come to the end, and it only remains to me to thank you all for having supported me by your presence in such numbers to-night, and to say to you in the words of Othello:

> " It gives me wonder great as my content,
> To see you here before me."

THE FOLLOWING EDITIONS OF OLD WORKS, IN ILLUSTRATION OF THE PAPER, WERE EXHIBITED BY THE AUTHOR.

1. JOHN WILKINS, (Bishop of Chester,) *Mathematicall Magick.* (First Edition.) Sm. 8vo. London, 1648.
2. —— *Ditto.* (Third Edition.) Sm. 8vo. London, 1680.
3. —— *Ditto.* (Fourth Edition.) Sm. 8vo. London, 1691.
4. Aulus Gellius, *Noctes Atticæ.* Folio. Paris, 1530.
5. —— *Ditto.* Sm. 8vo. Lyons, 1546.
6. —— *Ditto.* 12mo. (Elzevir.) Amsterdam, 1651.
7. Hero, of Alexandria. *Spiritalia.* (Commandinus Edition.) Sm. 4to. Urbino, 1575.
8. —— *Ditto.* (Aleotti Edition.) Sm. 4to. Ferrara, 1589.
9. —— *Ditto.* (Georgi Edition.) 4to. Urbino, 1592.
10. —— *Ditto.* (Aleotti Edition.) Sm. 4to. Amsterdam, 1680.
11. —— *De gli automati overo machine se movente.* Sm. 4to. Venice, 1589.
12. —— *Quatro theoremi aggiunti a gli artifitiosi Spiriti.* Sm. 4to. Ferrara, 1589.
13. John Bate, *The Mysteries of Nature and Art.* Sm. 4to. London, 1654.

14. Edward Somerset (Marquis of Worcester). *A Century of the Names and Scantlings of such Inventions, as at present I can call to mind.* 12mo. London, 1746.

15. Agostino Ramelli. *Le Diverse et artificiose Machine.* Folio. Paris, 1588.

16. Athanasius Kircher. *Magnes sive de Arte Magnetica.* Folio. Rome, 1641.

17. Vaucanson. *An Account of the Mechanism of Automaton or image playing on the German Flute.* 4to. London, 1742.

18. Peter van Musschenbroeck. *Introductio ad Philosophiam Naturalem.* 4to. Padua, 1768.

19. Jacques Ozanam. *Recréations Mathématiques et physiques.* 8vo. Paris, 1696.

20. Anonymous, (believed to be by Thomas Powell, D.D.) Humane Industry, or a History of most Manual Arts. Sm. 8vo. London, 1661.

21. Anonymous, (probably Professor Willis.) *An attempt to Analyse the Automaton Chess player of Mr. de Kempelen.* 8vo. London, 1821.

22. Cox's Museum. *Descriptive Catalogue of the Superb and Magnificent pieces of Mechanism and Jewellery in Cox's Museum.* Sm. 4to. London, 1772.

23. Henry Van Etten, *Mathematicall Recreations.* 12mo. London, 1633.

O. V.

A

BIBLIOGRAPHY

OF THE

PRIVATELY PRINTED OPUSCULA

Issued to the Members of the Sette of Odd Volumes.

" Books that can be held in the hand, and carried to the fireside, are the
best after all."—*Samuel Johnson.*

" The writings of the wise are the only riches our posterity cannot
squander."—*Charles Lamb.*

1. **B. Q.**

A Biographical and Bibliographical Fragment. 22 Pages. Presented
on November the 5th, 1880, by His Oddship C. W. H. WYMAN. 1st
Edition limited to 25 copies. (Subsequently enlarged to 50 copies.)

2. **Glossographia Anglicana.**

By the late J. TROTTER BROCKETT, F.S.A., London and New-
castle, author of "Glossary of North Country Words," to which is
prefixed a Biographical Sketch of the Author by FREDERICK
BLOOMER. (pp. 94.) Presented on July the 7th, 1882, by His Oddship BERNARD QUARITCH. Edition limited to 150 copies.

H

3. Ye Boke of Ye Odd Volumes,

from 1878 to 1883. Carefvlly *Compiled* and painsfvlly *Edited* by ye vnworthy Historiographer to ye Sette, *Brother* and *Vice-President* WILLIAM MORT THOMPSON, and produced by ye order and at ye charges of Hys Oddship ye President and Librarian of ye Sette, Bro. BERNARD QUARITCH. (pp. 136.) Presented on April the 13th, 1883, by his Oddship BERNARD QUARITCH.

Edition limited to 150 copies.

4. Love's Garland;

Or Posies for Rings, Hand-kerchers, & Gloves, and such pretty Tokens that Lovers send their Loves. London, 1674. A Reprint. And Ye Garland of Ye Odd Volumes. (pp. 102.) Presented on October the 12th, 1883, by Bro. JAMES ROBERTS BROWN.

Edition limited to 250 copies.

5. Queen Anne Musick.

A brief Accompt of ye genuine Article, those who performed ye same, and ye Masters in ye facultie. From 1702 to 1714. (pp. 40.) Presented on July the 13th, 1883, by Bro. BURNHAM W. HORNER.

Edition limited to 100 copies.

6. A Very Odd Dream.

Related by His Oddship W. M. THOMPSON, President of the Sette of Odd Volumes, at the Freemasons' Tavern, Great Queen Street, on June 1st, 1883. (pp. 26.) Presented on July the 13th, 1883, by His Oddship W. MORT THOMPSON. Edition limited to 250 copies.

7. Codex Chiromantiae.

Being a Compleate Manualle of ye Science and Arte of Expoundynge ye Past, ye Presente, ye Future, and ye Charactere, by ye Scrutinie of ye Hande, ye Gestures thereof, and ye Chirographie. *Codicillus I.* —CHIROGNOMY. (pp. 118.) Presented on November the 2nd, 1883, by Bro. ED. HERON-ALLEN. Edition limited to 133 copies.

8. Intaglio Engraving: Past and Present.

An Address by Bro. EDWARD RENTON, delivered at the Freemasons' Tavern, Great Queen Street, on December 5th, 1884. (pp. 74.) Presented to the Sette by His Oddship EDWARD F. WYMAN. Edition limited to 200 copies.

9. **The Rights, Duties, Obligations, and Advantages of Hospitality.**

An Address by Bro. CORNELIUS WALFORD, F.I.A., F.S.S., F.R. Hist. Soc., Barrister-at-Law, Master of the Rolls in the Sette of Odd Volumes, delivered at the Freemasons' Tavern, Great Queen Street, on Friday, February 5th, 1885. (pp. 72.) Presented to the Sette by His Oddship EDWARD F. WYMAN. Edition limited to 133 copies.

10. **"Pens, Ink, and Paper:" a Discourse upon Caligraphy.**

The Implements and Practice of Writing, both Ancient and Modern, with Curiosa, and an Appendix of famous English Penmen, by Bro. DANIEL W. KETTLE, F.R.G.S., Cosmographer; delivered at the Freemasons' Tavern, Great Queen Street, on Friday, November 6th, 1885. (pp. 104.) Presented to the Sette on January 8th, 1886, by Bro. DANIEL W. KETTLE. Edition limited to 233 copies.

11. **On Some of the Books for Children of the Last Century.**

With a few Words on the Philanthropic Publisher of St. Paul's Churchyard. A paper read at a Meeting of the Sette of Odd Volumes by Brother CHARLES WELSH, Chapman of the Sette, at the Freemasons' Tavern, on Friday, the 8th day of January, 1886. (pp. 108.) Presented to the Sette by Bro. CHARLES WELSH. Edition limited to 250 copies.

12. **Frost Fairs on the Thames.**

An Address by Bro. EDWARD WALFORD, M.A., Rhymer to the Sette of the Odd Volumes, delivered at Willis's Rooms, on Friday, December 3rd, 1886. (pp. 76.) Presented to the Sette by His Oddship GEORGE CLULOW. Edition limited to 133 copies.

13. **On Coloured Books for Children.**

By Bro. CHARLES WELSH, Chapman to the Sette. Read before the Sette, at Willis's Rooms, on Friday, the 6th May, 1887. With a Catalogue of the Books Exhibited. (pp. 60.) Presented to the Sette by Bro. JAMES ROBERTS BROWN. Edition limited to 255 copies.

14. A Short Sketch of Liturgical History and Literature.

Illustrated by Examples Manuscript and Printed. A Paper read at a Meeting of the Sette of Odd Volumes by Bro. BERNARD QUARITCH, Librarian and First President of the Sette, at Willis's Rooms, on Friday, June 10th, 1887. (pp. 86.) Presented to the Sette by Bro. BERNARD QUARITCH.

15. Cornelius Walford : In Memoriam.

By his Kinsman, EDWARD WALFORD, M.A., Rhymer to the Sette of Odd Volumes. Read before the Sette at Willis's Rooms, on Friday, October 21st, 1887. (pp. 60.) Presented to the Sette by Bro. EDWARD WALFORD, M.A. Edition limited to 255 copies.

16. The Sweating Sickness.

By FREDERICK H. GERVIS, M.R.C.S., Apothecary to the Sette of Odd Volumes, delivered at Willis's Rooms, on Friday, November 4th, 1887. (pp. 48.) Presented to the Sette by Bro. FRED. H. GERVIS. Edition limited to 133 copies.

17. New Year's Day in Japan.

By Bro. CHARLES HOLME, Pilgrim of the Sette of Odd Volumes. Read before the Sette at Willis's Rooms on Friday, January 6th, 1888. (pp. 46.) Presented to the Sette by Bro. CHARLES HOLME. Edition limited to 133 copies.

18. Ye Seconde Boke of Ye Odd Volumes,

from 1883 to 1888. Carefvlly *Compiled* and painsfvlly *Edited* by ye vnworthy Historiographer to ye Sette, Bro. WILLIAM MORT THOMPSON, and produced by ye order and at ye charges of ye Sette. (pp. 157.) Edition limited to 115 copies.

19. Repeats and Plagiarisms in Art, 1888.

By Bro. JAMES ORROCK, R.I., Connoisseur to the Sette of Odd Volumes. Read before the Sette at Willis's Rooms, St. James's, on Friday, January 4th, 1889. (pp. 33.) Presented to the Sette by Bro. JAMES ORROCK, R.I. Edition limited to 133 copies.

20. How Dreams Come True.

A Dramatic Sketch by Bro. J. TODHUNTER, Bard of the Sette of Odd Volumes. Performed at a Conversazione of the Sette at the Grosvenor Gallery, on Thursday, July 17th, 1890. (pp. 46.) Presented to the Sette by His Oddship Bro. CHARLES HOLME.
Edition limited to 600 copies.

21. The Drama in England during the last Three Centuries.

By Bro. WALTER HAMILTON, F.R.G.S., Parodist to the Sette of Odd Volumes. Read before the Sette at Limmer's Hotel, on Wednesday, January 8th, 1890. (pp. 80.) Presented to the Sette by Bro. WALTER HAMILTON. Edition limited to 201 copies.

22. Gilbert, of Colchester.

By Bro. SILVANUS P. THOMPSON, D.Sc., B.A., Magnetizer to the Sette of Odd Volumes. Read before the Sette at Limmer's Hotel, on Friday, July 4th, 1890. (pp. 63.) Presented to the Sette by Bro. SILVANUS P. THOMPSON. Edition limited to 249 copies.

23. Neglected Frescoes in Northern Italy.

By Bro. DOUGLAS H. GORDON, Remembrancer to the Sette of Odd Volumes. Read before the Sette at Limmer's Hotel, on Friday, December 6th, 1889. (pp. 48.) Presented to the Sette by Bro. DOUGLAS H. GORDON. Edition limited to 133 copies.

24. Recollections of Robert-Houdin.

By Bro. WILLIAM MANNING, Seer to the Sette of Odd Volumes. Delivered at a Meeting of the Sette held at Limmer's Hotel, on Friday, December 7th, 1890. (pp. 81.) Presented to the Sette by Bro. WILLIAM MANNING. Edition limited to 205 copies.

25. Scottish Witchcraft Trials.

By Bro. J. W. BRODIE INNES, Master of the Rolls to the Sette of Odd Volumes. Read before the Sette at a Meeting held at Limmer's Hotel, on Friday, November 7th, 1890. (pp. 66.) Presented to the Sette by Bro. ALDERMAN TYLER. Edition limited to 245 copies.

MISCELLANIES.

1. Inaugural Address

of His Oddship, W. M. THOMPSON, Fourth President of the Sette of Odd Volumes, delivered at the Freemasons' Tavern, Great Queen Street, on his taking office on April 13th, &c. (pp. 31.) Printed by order of Ye Sette, and issued on May the 4th, 1883.

Edition limited to 250 copies.

2. Codex Chiromantiae.

Appendix A. Dactylomancy, or Finger-ring Magic, Ancient, Mediæval, and Modern. (pp. 34.) Presented on October the 12th, 1883, by Bro. ED. HERON-ALLEN. Edition limited to 133 copies.

3. A President's Persiflage.

Spoken by His Oddship W. M. THOMPSON, Fourth President of the Sette of Odd Volumes, at the Freemasons' Tavern, Great Queen Street, at the Fifty-eighth Meeting of the Sette, on December 7th, 1883. (pp. 15.) Edition limited to 250 copies.

4. Inaugural Address

of His Oddship EDWARD F. WYMAN, Fifth President of the Sette of Odd Volumes, delivered at the Freemasons' Tavern, Great Queen Street, on his taking office, on April 4th, 1884, &c. (pp. 56.) Presented to the Sette by His Oddship EDWARD F. WYMAN.

Edition limited to 133 copies.

5. Musical London a Century Ago.

Compiled from the Raw Material, by Brother BURNHAM W. HORNER, F.R.S.L., F.R. Hist. S., Organist of the Sette of Odd Volumes, delivered at the Freemasons' Tavern, Great Queen Street, on June 6th, 1884. (pp. 32.) Presented to the Sette by His Oddship EDWARD F. WYMAN.   Edition limited to 133 copies.

6. The Unfinished Renaissance;

Or, Fifty Years of English Art. By Bro. GEORGE C. HAITÉ, Author of "Plant Studies," &c. Delivered at the Freemasons' Tavern, Friday, July 11th, 1884. (pp. 40.) Presented to the Sette by His Oddship EDWARD F. WYMAN.   Edition limited to 133 copies.

7. The Pre-Shakespearian Drama.

By Bro. FRANK IRESON. Delivered at the Freemasons' Tavern, Friday, January 2nd, 1885. (pp. 34.) Presented to the Sette by His Oddship EDWARD F. WYMAN.   Edition limited to 133 copies.

8. Inaugural Address

of His Oddship, Brother JAMES ROBERTS BROWN, Sixth President of the Sette of Odd Volumes, delivered at the Freemasons' Tavern, Great Queen Street, on his taking office, on April 17th, 1885, &c. (pp. 56.) Presented to the Sette by His Oddship JAMES ROBERTS BROWN.   Edition limited to 133 copies.

9. Catalogue of Works of Art

Exhibited at the Freemasons' Tavern, Great Queen Street, on Friday, July 11th, 1884. Lent by Members of the Sette of Odd Volumes. Presented to the Sette by His Oddship EDWARD F. WYMAN.   Edition limited to 255 copies.

10. Catalogue of Manuscripts and Early-Printed Books

Exhibited and Described by Bro. B. QUARITCH, the Librarian of the Sette of Odd Volumes, at the Freemasons' Tavern, Great Queen Street, June 5th, 1885. Presented to the Sette by His Oddship JAMES ROBERTS BROWN.   Edition limited to 255 copies.

11. Catalogue of Old Organ Music

Exhibited by Bro. BURNHAM W. HORNER, F.R.S.L., F.R.Hist.S., Organist of the Sette of Odd Volumes, at the Freemasons' Tavern, Great Queen Street, on Friday, February 5th, 1886. Presented to the Sette by His Oddship JAMES ROBERTS BROWN.

Edition limited to 133 copies.

12. Inaugural Address

of His Oddship Bro. GEORGE CLULOW, Seventh President of the Sette of Odd Volumes, delivered at the Freemasons' Tavern, Great Queen Street, on his taking office, on April 2nd, 1886, &c. (pp. 64.) Presented to the Sette by His Oddship GEORGE CLULOW.

Edition limited to 133 copies.

13. A Few Notes about Arabs.

By Bro. CHARLES HOLME, Pilgrim of the Sette of Odd Volumes. Read at a Meeting of the "Sette" at Willis's Rooms, on Friday, May 7th, 1886. (pp. 46.) Presented to the Sette of Odd Volumes by Bro. CHAS. HOLME.

Edition limited to 133 copies.

14. Account of the Great Learned Societies and Associations, and of the Chief Printing Clubs of Great Britain and Ireland

Delivered by Bro. BERNARD QUARITCH, Librarian of the Sette of Odd Volumes, at Willis's Rooms on Tuesday, June 8th, 1886. (pp. 66.) Presented to the Sette by His Oddship GEORGE CLULOW.

Edition limited to 255 copies.

15. Report of a Conversazione

Given at Willis's Rooms, King Street, St. James's, on Tuesday, June 8th, 1886, by his Oddship Bro. GEORGE CLULOW, *President;* with a summary of an Address on "LEARNED SOCIETIES AND PRINTING CLUBS," then delivered by Bro. BERNARD QUARITCH, *Librarian.* By Bro. W. M. THOMPSON, *Historiographer.* Presented to the Sette by His Oddship GEORGE CLULOW.

Edition limited to 255 copies.

16. Codex Chiromantiae.

Appendix B.—A DISCOURSE CONCERNING AUTOGRAPHS AND THEIR SIGNIFICATIONS. Spoken in valediction at Willis's Rooms, on October the 8th, 1886, by Bro. EDWARD HERON-ALLEN. (pp. 45.) Presented to the Sette by His Oddship GEORGE CLULOW.

Edition limited to 133 copies.

. **Inaugural Address**

of His Oddship ALFRED J. DAVIES, Eighth President of the Sette of Odd Volumes, delivered at Willis's Rooms, on his taking office on April 4th, 1887. (pp. 64.) Presented to the Sette by His Oddship ALFRED J. DAVIES.   Edition limited to 133 copies.

. **Inaugural Address**

of His Oddship Bro. T. C. VENABLES, Ninth President of the Sette of Odd Volumes, delivered at Willis's Rooms, on his taking office on April 6th, 1888. (pp. 54.) Presented to the Sette by His Oddship T. C. VENABLES.   Edition limited to 133 copies.

. **Ye Papyrus Roll-Scroll of Ye Sette of Odd Volumes.**

By Bro. J. BRODIE-INNES, Master of the Rolls to the Sette of Odd Volumes, delivered at Willis's Rooms, May 4th, 1888. (pp. 39.) Presented to the Sette by His Oddship T. C. VENABLES.
                     Edition limited to 133 copies.

. **Inaugural Address**

of His Oddship Bro. H. J. GORDON Ross, Tenth President of the Sette of Odd Volumes, delivered at Willis's Rooms, King Street, St. James's Square, on his taking office, April 5th, 1889.
                     Edition limited to 255 copies.

WORKS DEDICATED TO THE SETTE.

The Ancestry of the Violin.
London, 1882. EDWARD HERON-ALLEN.

An Odd Volume for Smokers.
London, 1889. WALTER HAMILTON.

The Blue Friars.
London, 1889. W. H. K. WRIGHT.

Quatrains.
London, 1892. W. WILSEY MARTIN.

Ꝑe Sette of Ꝋdd Columes.

Original Member. 1878. **BERNARD QUARITCH,** *Librarian,* 15, Picca-
dilly, W. (President, 1878, 1879, and
1882).

Original Member. 1878. **EDWARD RENTON,** *Herald,* 44, South Hill
Park, Hampstead, N.W. (Vice-Presi-
dent, 1880; Secretary, 1882).

Original Member. 1878. **W. MORT THOMPSON,** *Historiographer,* 16,
Carlyle Square, Chelsea, S.W. (Vice-
President, 1882; President, 1883).

Original Member. 1878. **CHARLES W. H. WYMAN,** *Typographer,* 103,
King Henry's Road, Primrose Hill,
N.W. (Vice-President, 1878 and 1879;
President, 1880).

Original Member. 1878. **EDWARD F. WYMAN,** *Treasurer,* 19, Blomfield
Road, Maida Vale, W. (Secretary, 1878
and 1879; President, 1884).

1878. **ALFRED J. DAVIES,** *Attorney-General,* Fair-
light, Uxbridge Road, Ealing, W. (Vice-
President, 1881; Secretary, 1884;
President, 1887).

1878. G. R. TYLER, Alderman, late High Sheriff of the City of London, *Stationer*, 17, Penywern Road, South Kensington, W. (Vice-President, 1886).

1879. T. C. VENABLES, *Antiquary*, 9, Marlborough Place, N.W. (President, 1888).

1879. JAMES ROBERTS BROWN, *Alchymist*, 44, Tregunter Road, South Kensington, W. (Secretary, 1880 ; Vice-President, 1883 ; President, 1885).

1880. BURNHAM W. HORNER, F.R.S.L., *Organist*, Matson Red House, Richmond Park, Richmond, S.W. (Vice-President, 1889).

1882. WILLIAM MURRELL, M.D., *Leech* (president), 17, Welbeck Street, Cavendish Square, W. (Secretary, 1883; Vice-President, 1885).

1883. HENRY GEORGE LILEY, *Art Director*, Radnor House, Radnor Place, Hyde Park, W.

1883. GEORGE CHARLES HAITÉ, F.L.S., *Art Critic*, Ormsby Lodge, The Avenue, Bedford Park, W. (Vice-President, 1887 ; President, 1891).

1883. EDWARD HERON-ALLEN, *Necromancer*, (Vice-President), 3, Northwick Terrace, N.W. (Secretary, 1885).

1883. WILFRID BALL, R. P. E., *Painter-Etcher*, 4, Albemarle Street, W. (Master of Ceremonies, 1890 ; Vice - President, 1891).

1884. DANIEL W. KETTLE, F.R.G.S., *Cosmographer*, Hayes Common, near Beckenham, Kent (Secretary, 1886).

1884. CHARLES WELSH, *Chapman*, The Poplars, Forest Lane, Walthamstow (Vice-President, 1888).

1886. CHARLES HOLME, F.L.S., *Pilgrim*, The Red House, Bexley Heath, Kent (Secretary, 1887 ; President, 1890).

1886. FREDK. H. GERVIS, M R.C.S., *Apothecary*, 1, Fellows Road, Haverstock Hill, N.W.

1887. JOHN W. BRODIE-INNES, *Master of the Rolls*, 14, Dublin Street, Edinburgh (Secretary, 1888).

1887. HENRY MOORE, A.R.A., *Ancient Mariner*, Collingham, Maresfield Gardens, N.W.

Supplemental Odd Volumes.

1887. JAMES ORROCK, R.I., *Connoisseur*, 48, Bedford Square, W.C.

1888. ALFRED EAST, R.I., *Landscape Painter*, 14, Adamson Road, Belsize Park, N.W.

1888. WALTER HAMILTON, *Parodist*, Keeper of the Archives, Ellarbee, Elms Road, Clapham Common, S.W.

1888. DOUGLAS H. GORDON, *Remembrancer*, (Master of Ceremonies), 41, Tedworth Square, S.W. (Secretary, 1889).

1888. ALEXANDER T. HOLLINGSWORTH, *Artificer*, 172, Sutherland Avenue, Maida Vale, W. (Vice-President, 1890).

1888. JOHN LANE, *Bibliographer*, 37, Southwick Street, Hyde Park, W. (Odd Councillor, 1891; Secretary, 1890; Master of Ceremonies, 1891).

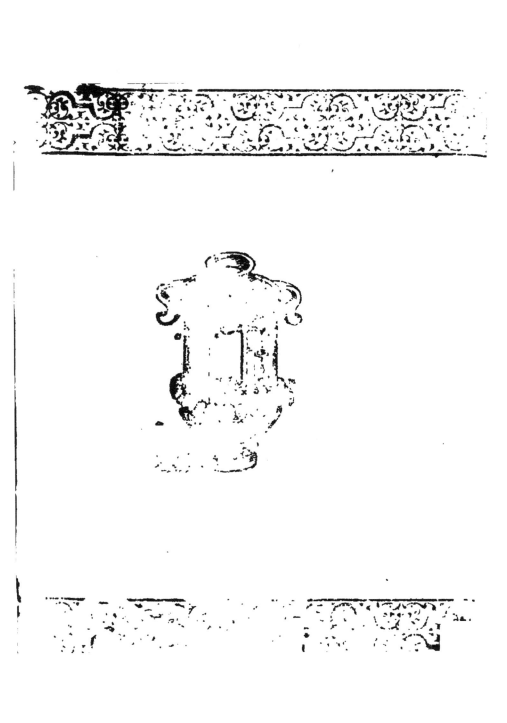

2